A-10 WARTHOG

Supreme in the Tank-Busting role

A-10 WARTHOG

Supreme in the Tank-Busting role

Robert F Dorr & Randy Jolly

OSPREY
AEROSPACE

Published in 1992 by
Osprey Publishing Limited
59 Grosvenor Street,
London W1X 9DA

© Robert F Dorr & Randy Jolly

ISBN 1 85532 195 5

Editor Dennis Baldry
Photography Randy Jolly (unless
otherwise credited)
Text Robert F Dorr
Page design Paul Kime
Printed in Hong Kong

Front cover Hey, Bob, this is Randy.
Yeah, man.
Hey, I want to let you in on one of
the best-kept secrets about the A-10.
Well, I don't know, I was talking to
the publisher. I don't think there's
anything to be gained by publishing
any *secrets* in our book.
I'm going to tell you anyway. You
know how everybody repeatedly
says how ugly the thing is?
Boy, I'll say. The only attack aircraft
vulnerable to bird strikes from the
rear. The Pig. The Hog. Some other
names we can't print. Bob, *that's*
what's they've been hiding.
Okay. I give up.
It's *not* ugly, Bob. You look at this
shot of New Orleans' 77-0240
banging through the air. It's not even
homely. Bob, that thing is downright
beautiful.
Randy, how come none of us ever
noticed that before?

Back cover CO's Warthog of the
23rd Fighter Wing feels for the
ground at England AFB, Louisiana.
This aircraft is also illustrated on
page 28

Title page The US Army term for
the illuminators glowing from this
Warthog are known as 'slime lights'
because of their yellow-greenish hue.
The US Air Force simply calls them
formation lights. This dusk silhouette
also accentuates the gap in the rudder
mass balance atop the vertical tail of
the A-10 Warthog as well as the
plain fact that the main landing gear
is only partially retractable. A-10 was
originally viewed as optimum for
daytime missions, but hi-tech in the
1990s has transformed it into a
nocturnal weapon as well

For a catalogue of all books published by Osprey Aerospace
please write to:

**The Marketing Department, Octopus Illustrated Books,
1st Floor, Michelin House, 81 Fulham Road, London SW3 6RB**

Captain Patrick Brian Olson
United States Air Force
21 April 1965 – 27 February 1991

A 1988 graduate of the United States Air Force Academy and a native of Washingtion, North Carolina, First Lieutenant Patrick B ('Oly') Olson, posthumously promoted to captain, was a forward air control or 'Nail FAC' pilot of a Fairchild OA-10 Thunderbolt II (77-0197, callsign NAIL 51) of the 23rd Tactical Air Support Squadron stationed at Davis–Monthan AFB, Arizona and flying from King Fahd Airport, Saudi Arabia during combat operations to liberate Kuwait. Operating from an FOL (forward operating location), Olson was hit by Iraqi gunfire for the second time in the Desert Storm conflict and was killed attempting to land his aircraft in bad weather. A cheerful young man with a relax manner that hid solid bedrock beneath,

Olson stands as a symbol for all those who fly and fight, and for those who willingly take the A-10 into harm's way. Olson knew, as some of us tend to forget, that liberty bears a price, that tyranny persists in our world, and that free men must remain prepared to employ force of arms. A charitable fund established in his honour assists deserving high school students in need of scholarships – the Patrick Olson Memorial fund, PO Box 1271, Washington, North Carolina 27889. As a symbol of all who fly and fight, of the indomitable spirit of those who choose the profession of aerial arms, this volume is dedicated to 'Oly' Olson – all honour to his name.

Introduction

The US Air Force has never liked the A-10. With the exception of those who work on, maintain, repair, arm, and *fly* the Fairchild-Republic A-10 Thunderbolt II, the venerable Warthog, most of the blue-suit bureaucracy and the Air Staff in particular have always given short shrift to the A-10 and to its mission. As this volume was going to press, it remained unclear how long the A-10 would remain in service. While drastic cutbacks in the lean post-Desert Storm era provided a convenient excuse to write *finis* to the A-10 story, the proven capability of the A-10 in the tank-killing and forward air control mission was so convincing, so overwhelming, that detractors were unlikely to get rid of the Hog nearly as soon as they might like.

We are not here to debate politics, especially not the politics of those who wear shade 84 blue and sit on the fourth floor, D-ring, of that funny five-sided building along the Potomac. Our message about the A-10 is for those who *do* work on, and fly, this remarkable machine and for those who appreciate it.

Some who appreciate the Hog are crew chiefs, pilots, commanders. Others of us are enthusiasts, model-builders, shutter bugs, and historians. There are enough of us. We are here to celebrate the Hog in the 1990s, and to say through the medium of pictures — most with Randy Jolly's skilled finger on the button — that the Warthog is still around. This pictorial essay is meant to show the mean, green tank-killer as it is today, and it is meant to entertain rather than to lecture. But Desert Storm has not left our memory and, on the preceding page, we make the serious point that liberty comes at a cost, and that free men must remain prepared to employ force of arms when need is.

The US Air Force deployed 144 A-10s to the Persian Gulf. Air superiority allowed innovative employment of the Warthog in a variety of roles. Primarily killing tanks on interdiction sorties, the A-10 proved its versatility as a daytime Scud-hunter in western Iraq, suppressing enemy air defences, attacking early-warning radars, and even recording two helicopter kills in the air — the only gun kills of the war. In combat, the A-10 flew just less than 8100 sorties and maintained a mission-capable rate of 95.7 per cent. The performance of the A-10 Warthog is no longer an issue for debate.

Any mistakes in this volume are the fault of the authors, but we would be remiss if we did not acknowledge the many who helped to make this work possible. Our thanks go out to many:

Among those who fly and fight: John Bradley, Herman (Hamster) Brunke, Ken (Bruiser) Brust, Jeffrey Fox, Bud Jackson, Dan Kuebler, Jeff (Moose) Musfeldt, John (Bear) Notestein, Denny Phelan, and Jim (Maggot) Preston.

In the world of public affairs, Don Black, John Carter, Mike Gannon, Susan Gentry, Bob Pease, Anna Pilutti, Clancy Preston, Cathy Reardon, Oscar Sierra, Kenneth Tackett, Dave Turner, Fran Tunstall, and Ron Woods.

Among the fraternity: Urs Bopp, David F Brown, Joseph G Handelman, Herbert D (Dan) Miller, Jr, the Gang at Roy's, Herman Sixma, Brian C Rogers, Norman E Taylor, Theodore Van Geffen, and Wally Van Winkle.

Robert F Dorr is a US Air Force veteran and author on military aviation whose by-line appears on some 35 volumes, and is co-author of *Tomcats Forever* in this series. Dorr has flown in seventy-six aircraft types. Bob lives in Oakton, Virginia with his wife Young Soon and sons Bobbie, 21, and Jerry, 17.

Randy Jolly is a defence and aerospace photographer whose camera work has appeared on the covers of *Newsweek, Aviation Week,* and many other international books and magazines. Jolly has logged hundreds of hour in military aircraft. Randy resides in Garland, Texas with his wife Andrea, daughter Erin, 18, and son Mac, 14.

Contents

Major Johnny Weaver removes his oxygen mask by undoing bayonet clasp on right side, following a mission in a 926th TFG A-10 Warthog on 10 January 1992. For those with an interest in close-up detail of canopy, cockpit, ejection seat, and other pilot-related doodads, the Air Force Reserve has kindly used stencilling in a couple of places to tell us which particular Warthog this is, namely 77-0208. The Warthog patch is used throughout the A-10 community

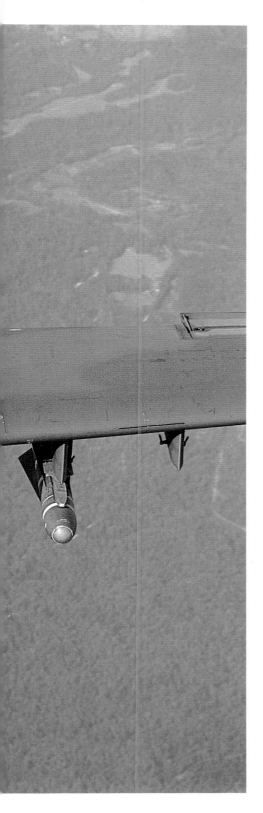

Pigs in space

Left This New Orleans A-10 Warthog is moving up to take a drink of jet engine fuel from a KC-135 tanker. Boom operators in the KC-135 say that the Hog is one of the easiest aircraft to refuel because of the far forward location of its receptacle for the 'flying boom' device. Wake turbulence from the tanker, while moderate, can be a problem until the A-10 pilot gains experience hanging on behind the larger aircraft. Many KC-135s of the new Air Mobility Command are scheduled to be modified for three-point refuelling, but the A-10 will be able to get three-at-a-time gas fillups only if – like some other Air Force tactical jets – it is modified to use the 'probe and drogue' system

Below Warthog 77-0202 of the 706th TFS/906th TFG, Air Force Reserve, forming on the tanker during a flight from home base at NAS New Orleans, Louisiana on 15 November 1991. White-painted guide marks on the tip of the nose in the shape of a widened letter 'I' are intended to guide the boom operator aboard a tanker, who must engage the Hog's refuelling receptacle

Two A-10s in formation. Note that while both carry the same ordnance load (two AIM-9M, two AGM-65C Maverick, on AN/ALQ-119(V), the armament is distributed on differing ordnance stations from one aeroplance to the other. These again are the supremely capable pilots of the Air Force Reserve's 442nd Tactical Fighter Wing headed up by Colonel John Bradley

Left Colonel John Bradley in 79-0111 slips ahead on the starboard front quarter of 79-0114 as the sprightly pair of Air Force Reserve A-10s march relentlessly over the Kansas flatlands. View from above provides a graphic examination of the A-10's unique lines, especially the twin turbofan engines mounted astride the rear fuselage. Intake located in dorsal position between the engines is a scoop for cooling of the very few vulnerable components in the rear fuselage

Above A fetchingly positive attitude seems to emanate from Colonel John Bradley, commander of the Air Force Reserve's 442nd Tactical Fighter Wing stationed at Richards-Gebaur AFB, Missouri. 79-0111 wears a KC tail code (not visible in this view) to denote its proximity to Kansas City and is normally flown by the wing's 303rd Tactical Fighter Squadron. While this appreciation of the Warthog was taking shape, the Reserve had not yet decided to follow the regular Air Force's practice of dropping 'Tactical' from its wing and squadron names

Left Passing beneath 79-0111 and 79-0114 in this two-ship formation, Richards-Gebaur Air Force Base is located 17 miles (27km) is named for 1st Lt John F Richards, killed 26 September 1918 in France on an artillery-spotting mission and Lt Col Arthur W Gebaur, Jr a F-84 Thunderjet pilot killed 29 August 1952 over North Korea on his 99th combat mission. In the world of those who fly and fight, it is possible to hold heroes like Richards and Gebaur in genuine reverence and still call up barracks humour to refer to the airbase in slang as Dicky Goober

Above Colonel John Bradley in 79-0111 and his wingman in 79-0114 show off the precision and discipline associated with the Air Force Reserve as they drive their 422nd TFW Warthogs past the camera. Even when rehearsing for war, A-10 pilots remain faithful to rigid safety rules, and even something as simple as a two-ship formation in level flight (accompanied by Jolly's camera ship) places a heavy demand for situational awareness and careful attention to every move

Above Colonel John Bradley bores a hole in the sky in 442nd TFW's 'boss bird,' 79-0111. Here's the exact skivvy on those seemingly innumerable hardpoints for ordnance:

The outboard under wing pylons, or stations 1 and 11 and 2 and 10 are stressed for loads of up to 1000 lb (454 kg);

The pylons immediately outboard of the undercarriage fairings (stations 3 and 9) carry up to 2500 lb (1134 kg);

The inboard underwing pylons (stations 4 and 8) and the fuselage outer pylons (stations 5 and 7) each carry up to 3500 lb (1587 kg);

The centreline pylon (station 6), although stressed for loads of up to 5000 lb (2268 kg) is not routinely used since it prevents the use of stations 5 and 7

— Stations 2, 10, 5, and 7 are removed here, as is often the case in high-threat situations. On the remaining hardpoints, Colonel Bradley carried (left to right), AN/ALQ-119(V) ECM pod on station 1; AGM-65 Maverick air-to-ground missiles on stations 3 and 9; and two AIM-9M Sidewinder heat-seeking air-to-air missiles on station 11. And if the preceding seems difficult to decipher, think of the challenges facing the dedicated enlisted personnel who load and arm these potent Warthogs!

Left and overleaf A-10 Warthog of the Air Force Reserve's 917th Tactical Fighter Wing, stationed at Barksdale AFB, Louisiana, closes in on Jolly's camera

Above 'Treble one' of the 442nd Tactical Fighter Wing, Richards-Gebaur AFB, Missouri, is heavily loaded and in a turn on a sunny day. Note that this A-10 carries two, rather than one, AIM-9M Sidewinder all-aspect heat-seeking missiles and an AN/ALQ-119(V) electronic countermeasures pod (the AN prefix indicating approval for joint-service use). AGM-65 Maverick air-to-ground missiles also hang from the wings, while fuselage hardpoints are vacant

Right A gaggle of 917th TFW Warthogs ploughs through the sky with the wing's 'sand and brown' airplane bringing up the rear. The Air Force Reservists maintain and fly these A-10s in superb condition. High visibility of the temporary, desert-style scheme in a non-desert environment is noteworthy

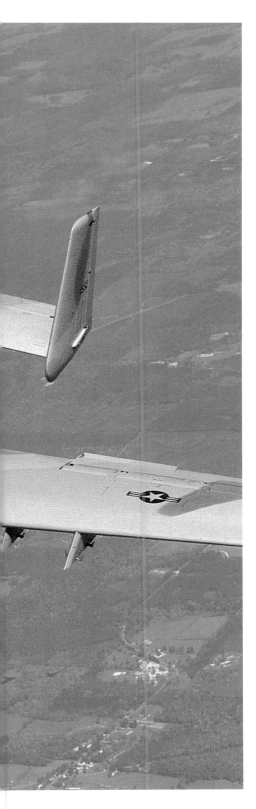

Left During Operation Desert Shield (just before the Persian Gulf war), the 917th Tactical fighter Wing at Barksdale AFB, Louisiana put a somewhat tacky-looking 'sand and brown' camouflage scheme on A-10 Warthog 78-0582, one of numerous experiments aimed at preparing for possible desert warfare. Quoth our shutterbug: 'In the air you could see it for miles. I believe they have scratched the idea of painting A-10s that colour'. The thinking was good, however, and many A-10 pilots in the Gulf complained that the most familiar lizard-green paint scheme made them dangerously visible to the enemy's Mark One Eyeball

Above Triple trouble. Flying in 'clean' condition without the payload which makes them so potent, New Orleans' 77-0271 (foreground), 77-0266 and 77-0268 pass in review during a 15 November 1991 sortie. This view accents the VHF (very high frequency) blade radio antennae protruding from the ventral area of the fuselage coincident with the engine exhaust line. The 706th TFS/926th TFG has operated the A-10 since June 1982, when it converted from the Cessna A-37B Dragonfly

Above The A-10 Warthog is not usually thought of as a creature of the night, but this nocturnal view does an excellent job of silhouetting the aeroplane's lines. Closest to camera is 79-0111, piloted by Lt Col Musfeldt commander of the 303rd Tactical Fighter Squadron. Second ship is 79-0093, followed by 79-0092, piloted by Captain Nate Stein and Captain Jim Preston. All belong to the 442nd Tactical Fighter Wing at Richards-Gebaur AFB, Missouri

Right Though they wear the WR tail code of the 81st Tactical Fighter Wing at RAF Bentwaters-Woodbridge, United Kingdom, this quartet of Warthogs is working out over the deserts of the American Southwest in 1987. Aircraft in foreground appears to be 77-0205, which later served with the New Orleans Air Reservists and, on 6 February 1991 shot down an Iraqi Bo.105 observation helicopter with its GAU-8/A cannon at the hands of Captain Robert R (Bob) Swain, 33, of Charlotte, North Carolina (*USAF*)

Above *Blue Thunder* is the nickname accompanying the snazzy nose art on this A-10 Warthog (apparently 81-0941) of the 81st Tactical Fighter Wing based at RAF Bentwaters-Woodbridge, United Kingdom in 1987. The white lightning bolt inside a blue fin cap flash, to say nothing of the 'blue' in the aircraft nickname, tells us that this ship belongs to the 91st Tactical Fighter Squadron, one of no fewer than four Thunderbolt II squadrons operated by this wing (*USAF*)

Right 358th FS A-10 Warthogs working out at Davis-Monthan. At the time of authors' visits to DM, Hog assignments were as follows:
602nd ACW:
   333rd FS (NF tailcode) OA-10 (former 22nd TASS)
   354th FS (NF) OA-10 (former 23rd TASS)
   355th FW (former 355th TTW and still a training wing):
   357th FS 'Dragons' (DM) A-10 (former 357th TFTS)
   358th FS 'Lobos' (DM) A-10 (former 358th TFTS)

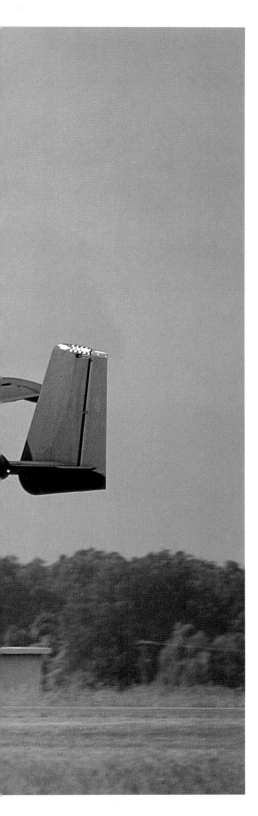

Left 'Boss bird' assigned to the commander of the 23rd Fighter Wing at England AFB, Louisiana (80-0191) springs into the sky wearing a unique blend of the fin cap colours of three squadrons assigned to the wing (74th, 75th, 76th). 75th FS has since been disbanded. Note 'highlighting' of tail code and wing designation with white-on-black on the Warthog's vertical tail

Above Night take off by 442nd TFW A-10 Warthogs from Richards-Gebaur demonstrates the forward retraction of the landing gear. On some aircraft types, lifting the gear by pulling it forward has created serious airflow and drag problems, but the Fairchild warplane seems to handle this arrangement with ease. Blade antenna standing at the centre spine of the aircraft is a UHF/TACAN antenna for communication and navigation

Pig power

Left Air Force Reserve A-10 pilot at Richards-Gebaur AFB, Missouri ('Dicky Goober' in jargon) climbs aboard 80-0110 in fading light preparing for a launch in late 1991. The 442nd Tactical Fighter Wing extended generous facilities to the authors. Their lizard-green Hogs were not merely porcine, but pristine. Note absence of scratch marks on air refuelling receptacle (forward of nose serial number)

Below These sharkteeth have a bite. Pre-Desert Storm portrait in 1989 shows the business end of a 76th TFS/23rd TFW Warthog as well as the rotary proboscis of the 30-mm GAU-8/A nose cannon. The gun is effective out to a range of 7000 yards, can penetrate the top of any main battle tank or armoured vehicle, and comes with a maximum of with 1175 rounds of ammunition. During Desert Storm, many potholes in paved roadways in Kuwait and Iraq were left behind after strafing runs by A-10s with this cannon

Loading belts of 30 mm depleted-uranium cannon cartridges aboard the A-10 Warthog sometimes calls for judicious use of musculature, but the men and women who arm and maintain this flying arsenal can, at least, derive a little benefit from hi-tech, including the ramp vehicle shown here. Despite the size and complexity of the cannon, which is so important that the Hog was virtually built around it, maintaining and arming the weapon is relatively easy to accomplish

This sequence of three views shows 'auto load' of the GAU-8/A 30 mm rotary cannon by highly skilled 442nd TFW armament experts. The cannon's ammunition drum holds 1350 rounds which are forced into its feed chute by the rotary action of the 'Gatling gun' helical chamber. Each cannon round is 11.4 in (290mm) in length and the complete API (armour-piercing incendiary) round weighs a formidable 2.05 lb (930 g)

Above An angry-faced AN/ALQ-119(V) ECM pod sneers from the outboard starboard-wing hardpoint of 80-0250 as hard-working armourers of the 23rd FW hand Maverick missiles on their large slow warplane. Ungraceful A-10 moves so slowly that it is widely reported to be the only jet aircraft vulnerable to bird strikes from behind. But the combination of Maverick and 30-mm GAU-8/A nose cannon with 1175 rounds of HEI (high explosive incendiary) ammunition means 'bad news' for enemy tanks

Left AN/ALQ-119(V) electronic countermeasures pod on an A-10 of the 23rd FW at England AFB, Louisiana in 1991. Given the increasing complexity of 1990s-era warfare and the growing importance of the electron on the battlefield, this gadget may be the most important of the numerous 'things under wings' not uncommonly found dangling from a Hog. ECM pod can jam and deceive radars being used to guide missiles which might otherwise ruin an A-10 driver's day

Above When the A-10 heads into Bad Guy country (formerly known as Indian country until the Native American Movement pretested), it must be prepared to deliver its bombs and missiles with a high degree of precision. Enhancing the bombing accuracy of the erstwhile Warthog is the Pave Penny laser guidance pod hanging from an off-centre left right-hand pylon beneath the cockpit. This system requires someone on the ground or in another aircraft to 'designate' a target with a laser beam, then pinpoints delivery of ordnance

Right A truly potent weapon for the A-10 Warthog is the AGM-65 Maverick missile which comes in electro-optical and television-guided versions. Although the Maverick is intended to be carried on a triple-station rack, as shown on this 23rd FW aircraft at England AFB, Louisiana, the inboard missile station is normally left empty, as shown, because at certain angles of attack the rocket blast can damage the Warthog's main landing gear assembly

Armourers at work in New Orleans. Hanging in the open position
at the front of the main undercarriage wheel housing is the A-10's pressure
fuel connection. An AGM-65 Maverick hangs just outboard of the main
landing gear, while these arming crews are adding a heavy dose of Mark 20
Rockeye CBU (cluster bomb units) farther outboard. The Hog can carry about
the same weight in air-to-ground ordnance as the B-17 Flying Fortress bomber
of World War 2

CBU-89 'Bomb, Cluster, Anti-tank,' seen here in non-opening training version, is externally similar to the Mark 20 Rockeye CBU package. The non-training variant is designed to fragment into 22 BLU-92 bomblets when dropped on a target. Note details of pylon attachment, which requires expert work by loaders

Above A look at some of the ordnance items carried by A-10. Blue training rounds in foreground are used to simulate the much larger Mark 82/83/84 series of gravity bombs, two of which occupy the centre of this picture between a pair of CBU-87 cluster bomb units

Right The shark's teeth on this 23rd Fighter Wing A-10 Warthog speak of the wing's historical link to the American Volunteer Group (AVG) and to the US 14th Air Force, the famed Flying Tigers of World War 2. Flying P-40s against Japanese Zeros, pilots at least did not have to worry about infrared seekers or surface-to-air missiles. In modern warfare, the threat has multiplied and the A-10 must be prepared to do exactly what this one *is* doing – drop flares to decoy an enemy's defences. A-10 also carries an ECM pod routinely in combat

Above Blue trim around the nose of this Mark 82 500 lb (227 kg) low-drag bomb identifies it as a training round, without explosive warhead. Warthog also carries typical air-to-air quiver of two AIM-9M Sidewinder missiles carried on LAU-105 carrier rails mounted to a dual rail adaptor (DRA). The Sidewinder load typically occupies either of the outboard ordnance hardpoints (stations 2, 10) while an ECM pod occupies the opposite station. Stateside Hogs carry the AN/ALQ-119(V)-15 ECM pod, while aircraft stationed overseas normally carry the AN/ALQ-131(V)

Right Wearing desert BDUs (battle dress utilities) which remain from their participation in the Persian Gulf war, armourers of the 23rd Fighter Wing carry a training version of the AIM-9M Sidewinder missile to be mounted aboard an A-10. The location is England AFB near Alexandria, Louisiana, which has been the 23rd FW's garrison in peacetime since its A-10 era began. In the aeroplane's only wartime deployment, the 23rd and other Warthog units flew from King Fahd airport, Saudi Arabia and from numerous FOLs (forward operating locations) during the fight against Iraq

Left Deceased T-55 main battle tank with a distinct puncture hole which may have been caused by an A-10's 30 mm GAU-8/A cannon, although the Warthog gun usually does not leave such a neat incision. Creating dead tanks (and more than a few potholes) was one of the major preoccupations of Warthog jocks who risked what is euphemistically called 'a high-threat environment' to carry the fight to the Iraqis during Operation Desert Storm (*Department of Defense*)

Above 1991 view of a 76 FS/23 FW Warthog making a low-level pass with GAU-8/A cannon ablaze. The 30 mm cannon employs a heavy, depleted uranium round which does not emit enough radioactivity to endanger those working on it, but does place the GAU-8/A in a special category for export purposes. Difficulty in obtaining clearance for the cannon round is thought to be the reason a planned A-10 purchase by Thailand fell through

Snout art

Close up of England-based Warthog

Left 23rd FW A-10s at England AFB, Louisiana in 1988

Above In silhouette, it might be mistaken for Mickey Mouse, but in broad daylight this portrait captures the essence of the A-10. Positioning the two 9065 lb (4109 kg) thrust General Electric TF34-GE-100 turbofan engines astride the rear fuselage, with exhaust sweeping over the B-25-style twin tail, may not have been a stroke of genius but it comes close. The Northrop A-9 aircraft which lost out to the A-10 for a production contract had more orthodox, internal engines

Above Sharkmouths galore. The location: England AFB, Louisiana. The outfit: the 23rd Fighter Wing, which in 1991 had the adjective 'Tactical ' dropped from its name. The occasion: a line-up of anti-tank attack aircraft. The teeth, gums, and tongue date back to General Claire Chennault's AVG (American Volunteer Group), the immortal Flying Tigers: every USAF unit since 1941 with a direct link to the Tigers has been authorized to wear these dentures

Right Major Johnny Weaver (left) and Lt Col Larry McCaskill compare notes in front of *Lucky Sun Dog* in a view which is particularly revealing of the retractable ladder arrangement on the high-seated Hog. Green thing on Larry's head is not for St Patrick's Day but is a straightforward skull cap employed to ease the 'fit' of the HGU-55P pilot's helmet: a poor fit can cause 'hot flashes' around the head which can be exceedingly painful and distracting. In old days, pilots wore their parachutes and carried them to and from ejection seats; nowadays the chute remains part of the ejection seat, leaving the pilots without a heavy load on their backs

Above 77-0240 *Lucky Sun Dog* with the 1990s version of a Vargas girl inside
the nose panel's ladder access door and a tank-killing lobster on the nose, may
belong to Major Rasar, but we think that's Major Johnny Weaver finishing a
flight in the Hog by glancing over the price you have to pay for getting to fly –
paperwork. The location is Alvin Callender Field in Belle Chasse, Louisiana, and
the aeroplane is an NO-coded A-10 belonging to the Air Force Reservists who
hail from in and around New Orleans

Inside panel of ladder door of this OA-10 seems to be telling the reader that it's not a good 'deal' to stake your hand against the 23rd Tactical Air Support Squadron, though the squadron had been redesignated 358th Fighter Squadron at Davis-Monthan by the time this view was taken. Not uncommonly, the aircraft data block, intended to tell us *which* Hog this is, contains an error. With the final digit of the serial covered by the ladder access door, we know that this aeroplane's serial number falls somewhere in the 76-0540 to 76-0549 range (covering the 540th through 549th aircraft or missiles ordered by the USAF in fiscal year 1976). Somehow, however, '76' became '7A' on the data plate. OA-10 units at this time were the 333rd FS and 354th FS (former 22nd and 23rd TASS), both with NF tail code

Being an A-10 pilot isn't immediately obvious as the most glamorous line of work in the world, but the men who fly the Warthog seem almost to glory in the ugliness of their flying machine and the difficulty of their mission. Shortly after this A-10 pilot got ready to go at England AFB, Louisiana in mid-1991, the 23rd FW (formerly the 23rd TFW) lost one of its squadrons when the 7th FS (formerly 75th TFS), whose tigershark emblem appears here, was deactivated on 1 October 1991

Warthog 77-0266 *Iraqi Nightmare* of the 926th Tactical Fighter Group in New Orleans on 10 January 1992. Months after Desert Storm, the 926th TFG was the only outfit in the A-10 still wearing nose art from the war. Major Will Shepard is attired in typical haberdashery for today's well-dressed Hog Driver

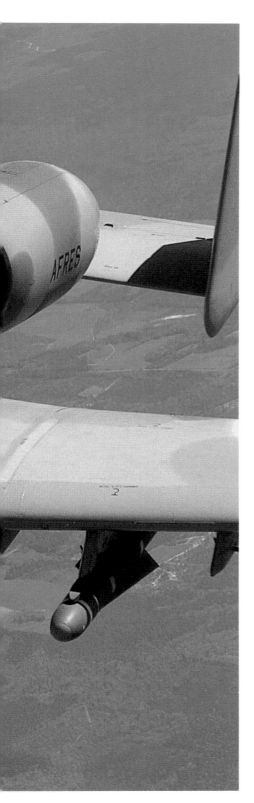

Desert Warthog? Nope. The 'sand and brown' camouflage applied to this Barksdale-based A-10 was not adopted by the Air Force. Aircraft 78-0582 is operated by the 917th Tactical Fighter Wing of the US Air Force Reserve (see pages 90-93)

Above Warthog 77-0227 of the 926th Tactical Fighter Group wears warpaint
on the ramp at New Orleans on 10 January 1992. The AGM-65 Maverick
missile is prominently displayed in this nose-art caricature. Though the A-10
was built around its General Electric GAU-8/A 30-mm rotary cannon for the
anti-tank role, the TV- and electro-optical Maverick (used, respectively, on day
and night missions) has proven effective against tanks and a variety of other
targets, including Scud missile launchers

Right Panicked Bad Guy in the saddle flees in haste from camel-seeking missile
in this 10 January 1992 portrayal of surviving Gulf War nose art on Fairchild
A-10 Thunderbolt II (77-0271) of the Air Force Reserve's 706th TFS/926th
TFG at New Orleans. 'Kills' accomplished in Bad Guy Land adorn the nose of
this veteran Hog and demonstrate that the aircraft, although it has never been
popular in the Pentagon bureaucracy, is far more versatile than its original tank-
killing mission might suggest

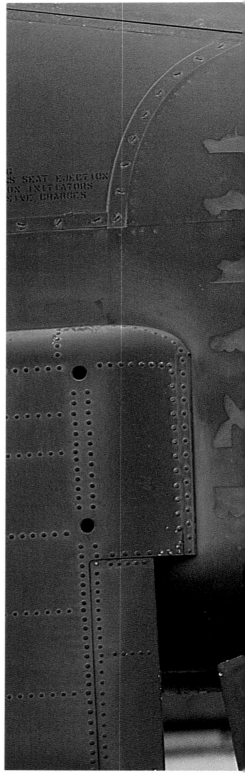

In good taste, but posing the risk of getting our minds away from aeroplanes, nose art on 77-0274 is another reminder of combat in the Persian Gulf. On 10 January 1992, this ship was part of the 706th TFS/926th TFG at New Orleans and was back to carrying out peacetime operations. Aerodynamic contours of *New Orleans Lady* may or may not be typical of pulchritude in that great southern city, but nose art is good for morale and commanders ought to consider allowing more of it in other units

Stephanie Ann, the Bayou Babe, adorns the nose of 77-0531 (a bayou is a marshy inlet or outlet of a river, and plenty of them straddle the approaches to the delta of the Mississippi River). Note that 'Mouse' (the crewchief?) is immortalized separately on this Hog's nose wheel door. Stephanie's stylized Maverick missile seems to have toted up an especially impressive number of non-functioning Iraqi trucks

Right In an armed service which has made dramatic strides toward a smoke-free environment, this nose art reproduces the likeness of a famous spokesman for a tobacco product. It was, perhaps, inevitable when the 926th TFG found itself heading off to fight in a region populated with numerous camels. Like other jets, 77-0255 wears its roster of 'kills' of Iraqi targets in this 10 January 1992 portrait

Above *Desert Rose* is a mean, green tank-killing, alias Fairchild A-10 Thunderbolt II 77-0273, viewed from a three-quarter angle which shows the ACES II ejection seat covered over but gives us a glimpse of the pilot's HUD (head-up display). The A-10 was designed for visual navigation and target acquisition and the avionics suite was kept simple, the HUD providing the Hog Driver with fairly basic symbology to enable him to keep his head up while fighting at low level in a high-threat zone

Above Seriously in need of dental work before periodontal infection becomes incurable, *Cresent City Desert Darlyn* is 77-0268 belonging to the New Orleans mob and appears fully ready to exhale some bad breath from its GAU-8/A. Note earphones and wire which permit the Hog's pilot to talk to the ramp crew member helping to prepare the A-10 for flight. Bright-red safety garment worn by the latter reflects simple common sense and the paperwork is, alas, more and more a fact of life in the Air Force

Right Already high off the ground, Lt Col Larry McCaskill of the 926th Tactical Fighter Group appears to be just about ready to roll out and venture into the sky in war veteran A-10 Warthog 77-0240. The date is 10 January 1992 and after coming home from the Middle East, the erstwhile Hog is still adorned in the 'Europe One' or lizard-green paint scheme which is standard and which, in some environs, makes the A-10 about as visible as a sore thumb

Hog squadrons

Left Haberdashery for the trendy A-10 jock will never make the pages of *Gentlemen's Quarterly*, but it does the job. This Nellis-based 57th Fighter Weapons Wing pilot wears HGU-55/P helmet (of much lighter weight and greater comfort than its predecessors), sun visor, T-10 oxygen mask, flight coveralls, G-suit (not visible), and gloves which are always worn during takeoff and landing because of fire hazard. Upper portion of ACES II ejection seat is revealed to good advantage in this profile

Below A-10 pilot at work in 75-0280, one of the earliest Hogs still serving. The badge on the pilot's right breast is the familiar emblem of Tactical Air Command, which was scheduled to be merged into the Air Combat Command on 1 July 1992. Badge on left shoulder (including a laid-over triangle enclosing a sword pointing toward the aircraft nose) belongs to the Davis-Monthan-based 355th Fighter Wing, which includes this pilot's squadron, the 358th FS, known as the 'Lobos'. Necessary flight information, the inevitable paperwork, is conveniently wedged between windshield and head-up console

Left Details of the Warthog's rather straightforward landing gear arrangement, including off-centre nose wheel and main wheels which retract into leading-edge fairings, are evident in this mid-1991 view of an aircraft of the 358th Fighter Squadron, 'Lobos', lifting off from Davis-Monthan Air Force Base near Tucson, Arizona. This view also accentuates the enormous size of the Warthog, which is one of the largest combat aircraft flown by a single pilot

Above 358th Fighter Group A-10 Warthog at Davis-Monthan in late 1991

Based at Davis-Monthan AFB, Arizona, the 357th Fighter Squadron (formerly a Tactical Fighter Training Squadron, and still assigned a training mission) is known as the Dragons and wears this distinctive badge

This A-10's fin's black band with a light grey wolf head identifies the 358th Fighter Squadron at Davis-Monthan AFB, Arizona, known until late 1991 as the 358th Tactical Fighter Training Squadron and – then and now – employed as an RTU (replacement training unit) for the Warthog type. 75-0306 is one of the earliest A-10s in the US Air Force inventory

358th Fighter Squadron A-10 Warthog landing at Davis-Monthan in late 1991

Left A-10 Warthog 77-0306 taxies at Davis-Monthan AFB, Arizona in late 1991. The Hog belongs to the 358th Fighter Squadron at D-M, just outside Tucson. 357th and 358th FS were the two squadrons assigned to the 355th FW at the end of 1991 and retained the training function they had always had when known earlier as Tactical Fighter Training Squadrons

Above The sunshine in the American Southwest is always a joy, and nowhere does it shine brighter than at Nellis AFB, Nevada, where a band of Maryland-based A-10 pilots visited from less sunny climes, showed their stuff, and walked away with top 'Gunsmoke' 1991 honours. 78-0602, being readied by hard-working ground crewmen, is yet another of the Warthogs from Baltimore which performed so well out in the desert near Las Vegas

Above Winning top honours at the 1991 'Gunsmoke' meet at Nellis AFB near Las Vegas simply *had* to be an uplifting experience for the dedicated Air Guardsmen of Maryland's 175th TFG, who have been largely overlooked in the literature of the A-10 and who did not have an opportunity to serve in Desert Shield/Storm. 79-0705 belongs to the group's 174th TFS and seems to be about ready to head out to the range and unleash some ordnance

Right Although the A-10 Warthog is loved by those who work on it and fly it (and disliked by Iraqis in uniform), the US Air Force's bureaucracy has never truly accepted the aircraft and has been trying to get rid of it ever since it first flew. It must have been a dreadful blow to the paper-pushing blue suiters when A-10s walked away with the Tactical Air Command 'Gunsmoke' 1991 competition. Aircraft 79-0704 belongs to the winners of the 'Gunsmoke' event, the 104th Tactical Fighter Squadron, 175th Tactical Fighter Group, Maryland Air National Guard, stationed at Glenn L Martin Airport in Baltimore. When this volume was being prepared, the active-duty US Air Force had dropped the adjective 'Tactical' from its fighter wings and squadrons, but reserve components were still using it

Left At 'Gunsmoke' 1991, where fighter-bomber folk convene to challenge each other's skills, the engine cover on this Warthog is a circular version of the flag of Maryland, one of the original thirteen American colonies and also location of the Hagerstown plant where A-10s were assembled. The Maryland Air National Guard's 175th Tactical Fighter Group has flown, in succession, F-86H Sabre, A-37B Dragonfly, and A-10 Warthog. When not competing at Nellis, the Maryland A-10s live at Glenn L Martin Airport near Baltimore

Above Slated for demobilization in the 'downsized' US Air Force of the 1990s, the 354th Fighter Wing at Myrtle Beach, South Carolina, has a checkered history which includes flying the F-100D Super Sabre, A-7D, and A-10 Warthog. MB-coded Hogs from Myrtle Beach went to war in the Persian Gulf along with other units. Here, the Myrtle crowd participates in 'Gunsmoke 1991', the semi-annual Tactical Air Command/Air Combat Command air-to-ground contest held at Nellis AFB, Nevada – an event in which other Warthogs from the Maryland Air National Guard took top honours

Above Rapidly fading in the Arizona sun, the emblem of the 23rd Tactical Air Support Squadron, the 'Nail FACs' is also fading into the history books. The squadron distinguished itself in Vietnam in the OV-10 Bronco and in the Persian Gulf in the OA-10 Warthog. By the time of this late 1991 view at David-Monthan AFB, however, the unit's identity had been shifted to the 354th Fighter Squadron. The squadron remained part of the 602nd Tactical Air Control Wing at press time, although speculation was rife that it would be absorbed by the 355th Fighter Wing

Right In late 1991, the 'Nail FACs' still wore their time-honoured NF tail code, even though the squadron had changed its designation from 23rd TASS to 354th FS. Although the OA-10 Warthog remained very much a part of Air Force plans in the forward air control mission, the passing of distinct FAC squadrons was a fact of life in the leaner times of the early 1990s. When this volume was being prepared, the FACs still had a distinct *wing* at Davis-Monthan, the 333rd FS and 354th FS (former 22nd and 23rd TASS) reporting to the 602nd Tactical Air Control Wing

Above Apart from the obvious resemblance with large wing, twin engines, and twin tail, the Fairchild A-10 Thunderbolt II is roughly the size of a B-25 Mitchell bomber of World War 2. Lest any doubt persist that the Warthog is downright awesome, Jolly's view of the aircraft from below its massive wing is submitted as evidence. 79-0210 belongs to the 74th Fighter Squadron, part of the 23rd Fighter Wing

Left 354th Fighter Wing A-10s lined up at Myrtle Beach during Gunsmoke 1991

Above The US Air Force system of unit identifiers, better known as tail codes, began in late-1966 when two-letter designators were assigned to tactical squadrons. In June 1972, the system was changed so that, in most cases, the tail codes belonged to wings rather than squadrons. In the case of the Air Reservists of the 926th TFG who fly the Fairchild A-10 Thunderbolt II, the code belongs to the *group*. While the word 'No' probably connotes what Saddam Hussein might say if asked whether he wants to see an A-10 again, the NO code on 77-0269 actually signifies the group's proximity to New Orleans

Left The 74th Fighter Squadron (part of the 23rd Fighter Wing) wears a white field and white flash as its fin cap colour. The EL code signifies England Air Force Base, near Alexandria, Louisiana, which in past times has been home to F-100 Super Sabres and A-7 Corsair IIs. The Fairchild A-10 Thunderbolt II, rarely called anything but the Warthog or the Pig, has been a familiar sight around England AFB since the early 1980s

Above The BD tail code is shorthand for Barksdale, the air base near Bossier City, Louisiana which is better-known for its large fleet of B-52 Stratofortresses and KC-10 Extenders. The Air Force Reservists of the 917th TFW operate their Hogs from one corner of the base (not near Hangar 10, where 78-0539 is sitting). The metallic blue fin cap with gold bottom border identifies 539 as belonging to B Flight of the wing's 46th TFTS, the RTU (replacement training unit) for Reserve A-10 forces. The wing also has the 47th TFS, which has an operational rather than a training mission

Right *Derrière* view of 917 TFW Barksdale A-10. Note the raised canopy and squared shape (between engines) of the air conditioning system intake and exhaust duct. Centre 'teat' at rear of fuselage is a tail navigation light, flanked by two formation running lights. Nosewheel is, of course, off-centre and inboard wing stores pylon is used as a 'hanging place' by pilots who often will leave a harness or even a G-suit hung there when not yet in the cockpit flying the aircraft

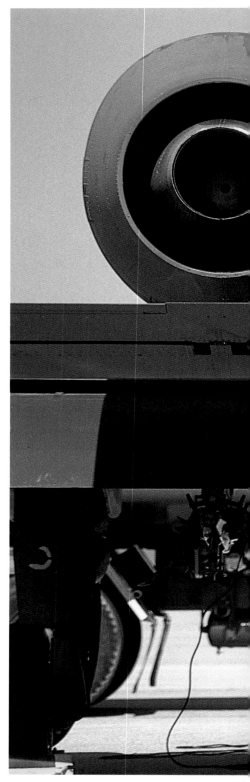

Walk-around examination of the 'sand and brown' camouflage on an A-10 Thunderbolt II (78-0582) belonging to the 917th Tactical Fighter Wing of the US Air Force Reserve at Barksdale AFB, Louisiana. A paint scheme similar to this was tried on very early A-10s in the mid-1970s during joint-service air-ground operations in California – and was rejected. This was another attempt and was not picked up by the Air Force, either

Walk-around completed, let's enjoy the sight of 78-0582 taxiing out for take-off

Above Details of the tails are apparent in this line-up at Nellis AFB of WA-coded A-10 Thunderbolts belonging to the 57th Fighter Weapons Wing, including 'boss bird' 79-0170 with it highlighted code and serials. The 57th FWW operates most of the tactical warplanes in US Air Force inventory to develop techniques, tactics, and aircraft improvements, and includes some of the most experienced Warthog pilots and support people in the service

Left The Warthog and the Aardvark have plenty in common. Neither is beautiful. Neither ever yanked and banked with a MiG-29. Neither has ever been immortalized in a Tom Cruise offering from Hollywood. All they do is carry ordnance, fly, fight, and kill tanks. The F-111 in foreground in this Nellis AFB portrait was not, of course, designed as a tank-killer but in the Persian Gulf managed to provide some serious competition to the A-10 in the latter's primary role. Both of them also joined in the campaign against Saddam's Scud missile launchers. Here at Nellis, A-10 and F-111 look as if they were made for each other

Above A-10 82-0646 was one of the last Warthogs assembled at Fairchild's Hagerstown, Maryland facility (now the offices of a credit-card company) and is seen, in this August 1989 view, with highlighted WR tail code of the 81st Tactical Fighter Wing at RAF Bentwaters-Woodbridge, United Kingdom. At one time, the 81st was the largest fighter wing the US Air Force ever had, with 110 or more Hogs on charge. With the closing of the RAF station, it's no more Warthogs in the wilds of East Anglia (*Herman Sixma/IAAP*)

Left A-10 80-0209 wears the checkered fin cap colours of the 75th Fighter Squadron (deactivated in October 1991), part of the 23rd Fighter Wing at England AFB, Louisiana and is shooting landings in this mid-1991 portrait. The lizard-green 'Europe One' camouflage paint scheme has been *de rigueur* for the A-10 Warthog since early in its service career but it was *not* what pilots wanted in a Mid-eastern desert environment where the 'Hog' stood out like a sore thumb

Left Warthog 79-0182 wears the star-spangled red fin cap colour of the 76th Fighter Squadron, a part of the 23rd Fighter Wing at England AFB, Louisiana. Squadron badge also appears beneath blade antenna at mid-fuselage. Note also AIM-9M Sidewinder and AGM-65 Maverick cargo hanging from this lightly-loaded Warthog, which is rehearsing the art of low-level strike warfare at its home base in mid-1991

Above Every one of those little red banners furled in the wind beneath the taxiing Warthog will be removed at 'last chance' near runway's end before the veritable Hog leaps into the sky. Taxiing out in 1991, this 75 FS/23 FW A-10 at England AFB, Louisiana imparts some of the brute force and enormous power of this attack aircraft, both of which are deemed by warfighting planners as far more pertinent than gracefulness

Above We don't usually think of corn-growing Kansas as part of the American Southwest, but a sunrise (or sunset) like this one has that special quality unique to a vast nation with Wide Open Spaces. All of the open space out in that vast heartland also provides room for Warthog pilots to rehearse their trade under realistic conditions. This A-10 attack aircraft of Colonel Bradley's 422nd TFW may be merely a silhouette now, but when it's time to go to work the Hog will carry bombs and missiles on real-life mission profiles in those vast expanses

Left Hog driver of the 358th Fighter Squadron at Davis-Monthan AFB, Arizona in November 1991, in silhouette. Outlines of helmet, visor, oxygen mask, HUD (head-up display), and ACES II ejection seat are all evident in this portrait. Vic Caputo, a radio talk-show host in Tucson, says that, 'We all know and recognize the Warthog in this community, and we all love it'. Davis-Monthan has traditionally fielded RTU (replacement training unit), FAC (forward air control), squadrons in its 602nd ACW and 355th FW respectively

... and pigs might fly

Left Back form Desert Storm, wearing the symbols of battle, and receiving attention from a maintenance technician, this A-10 Warthog (77-0272), pristine but for its scratchy refuelling receptacle, belongs to the Air Force Reserve's 706th Tactical Fighter Squadron, 906th Tactical Fighter Group, home-ported at NAS New Orleans, Louisiana. The aircraft is located in the phase hangar on 10 January 1992. The Group flew 900 combat sorties in the Persian Gulf conflict and racked up an impressive score of deceased Iraqi tanks, armoured vehicles, Scud launchers, and other impediments

Below Once the shooting had ceased, the authorities were not enthusiastic about distinctive markings on the aircraft which won the contest with Saddam Hussein. Aircraft 77-0268, at least, preserves the dentures of the venerable Hog and appears ready to set forth and be mean again, if it has to. The Air Reservists of the 706th TFS/926th TFG went to the conflict from NAS New Orleans, also known as Alvin Callender Field, located in Belle Chasse on the west bank of the Mississippi River just 10 miles (16 km) from the world-famous French Quarter

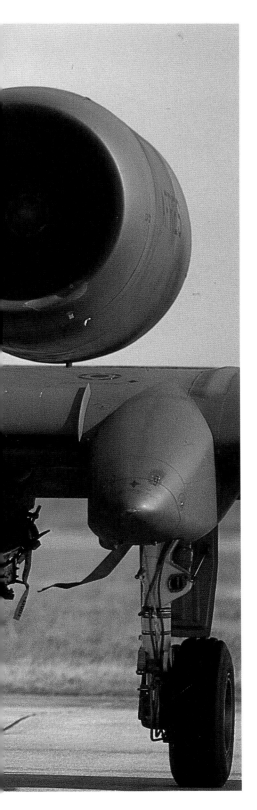

Left Many months after the war, New Orleans' 77-0269 wears nose art to commemorate the victory in the Gulf. This frontal view accentuates the asymetrical shape of the A-10 Warthog which causes some to opine that the craft is less than a thing of beauty. The cannon is canted two degrees downward from the fuselage line (but centred), the nosewheel is entirely off-centre, and the Pave Penny targeting device hangs from the right side of the nose. Coming on like this, the Hog exudes size, strength, and power

Above In blazing sunlight at New Orleans on 10 January 1992, A-10 Warthog 77-0269 pauses at 'last chance' for a look-over by the Air Reserve's capable armourers before lifting into the sky. AGM-65 Maverick hanging under left wing comes in electro-optical and laser-guided versions while AIM-9M Sidewinder missile to starboard is insurance against MiG trouble. Pilot of this 926th TFG jet is almost certainly fully prepared to 'shoot down what's up and shoot up what's down'

These pages and overleaf Technicians remove the mighty GAU-8/A cannon from the fuselage of A-10 Warthog 77-0273 in New Orleans on 10 January 1992. The biggest gun carried by any combat aircraft, the GAU-8/A has seven barrels rather than the six employed by the ubiquitous M61A1 Vulcan 20 mm cannon. Each barrel is specified to have a 21,000-round firing life. Not exactly 'modular' and not all that easy to install or remove, the GAU-8/A places strong demands on the people who work on it

Gently does it. Technicians toil away as the A-10's Avenger cannon parts
company with the airframe. In-service support for the Warthog is now the
responsibility of Grumman Aircraft Systems

Above On 19 January 1992, Major Will Shepard of the 926th TFG gets a
helping hand as he begins the descent from 77-0269 following a flight. Fatigue
uniforms worn here by ground crew members are known as BDUs (battle dress
utilities) and are now standard among all branches of the US armed forces, each
of which formerly had its own work attire

Right The McDonnell Douglas ACES II (Advanced Concept Ejection Seat)
provided to the pilot of the Fairchild A-10 Thunderbolt II has a superb record of
saving life in the event of emergency (and in combat). By definition, however,
no ejection seat is easy to put into, or take out of, an aircraft, so these
maintenance people of the 23rd FW may be sweating a little as they pursue
their delicate task at England AFB, Louisiana

23rd FW commander's 'Boss Bird' rolling at England AFB, Louisiana in mid-1991

113

Left Flight line of A-10 Warthog fighter-bombers at Davis-Monthan AFB, Arizona. Aircraft in foreground belongs to 358th Fighter Squadron. The importance of well-choreographed work efforts on the ramp is suggested by the clutter and crowding seen here. Safety is always stressed, ladders on a flight line area are highly unpopular, and everyone strives for the neatness which is also a part of this panorama

Above Though it wasn't posed for the purpose, this downward look at 79-0114 on the flight line at Richards-Gebaur AFB, Missouri (in 1991) shows some of the paraphernalia which goes along with repairing, maintaining, and pre-flighting the Fairchild A-10 Thunderbolt II. Ladders are despised by flight line crews and pilots alike but must occasionally be used on an aircraft which stands so high off the ground. Fire bottle, power cart, ammunition carrier, even the innocuous-looking Mark One Waste Can, are all part of the job of 'keeping 'em flying' whilst also avoiding FOD (foreign object damage) on a busy airfield ramp

Above A pair of A-10s rolling at dusk

Left Without crew chiefs like the technical sergeant strapping this Hog Jock into his ACES II ejection seat, no combat aircraft would even find its way out of the parking ramp, let alone to the target. Each crew chief is assigned a specific airframe for which he has responsibility, and treats it just as he would his personal property. Because the company which manufactured the A-10 no longer exists, it is becoming increasingly difficult to support, but crew chiefs, maintenance people and armourers speak highly of the aircraft and want to keep it in inventory

Hog history

When the USAF launched its A-X attack aircraft programme in the late 1960s, several manufacturers submitted proposals. Two were selected for a 'fly-off' competition, the Northrop A-9 and the Fairchild A-10. Both made their first flights in 1972. The Northrop YA-9A was an excellent aircraft but it lost out to the YA-10A service-test version of the Warthog. Today, the tank-killer that 'might have been' can be seen in the form of this surviving YA-9A (71-1367) on outdoor display at Castle AFB, California (*Larry Norris*)

The US Air Force did not originally
intend to paint the Warthog in its
now familiar 'Europe One' green
scheme. Early in its developmental
period, a number of experimental
paint designs were tried, including
this mottled battlefield grey seen on
A-10 73-1669 at Edwards AFB,
California in about 1977.
Developmental Warthogs in this
series had significant internal
differences from production aircraft,
and were of little use when the time
came to move them into operational
squadrons (*Robert Loftberg*)

Left When first placed into operational service, A-10 Warthogs wore the grey camouflage scheme illustrated here. This aircraft (77-0181) belongs to the 354th Tactical Fighter Wing (MB tail code) at Myrtle Beach AFB, South Carolina, and is shown during an Armed Forces Day visit to Andrews AFB, Maryland on 20 May 1978 (*Robert F Dorr*)

Above Early grey tactical paint scheme associated with the A-10 Warthog is worn on this WR-coded example from the 81st Tactical Fighter Wing at RAF Bentwaters-Woodbridge, England. This A-10 Warthog (77-0242) is taxiing during a visit to Soesterburg, Holland, on 27 June 1979 (*Urs Bopp*)

Aircraft 76-0530, one of the oldest A-10 Warthogs remaining in inventory, is a rare 'one of a kind' treasure for those who feast on oddball aircraft colours. The two-tone grey scheme of this Barksdale-based BD-coded Hog of the Air Force Reserve's 917th Tactical Fighter Wing evokes memories of experimental A-10 paint schemes tried in the mid-1970s but was being worn by the venerable 76-0530 in late 1991. At one time, such a scheme was under consideration for those A-10s assigned to the forward air control mission, which are designated OA-10 but have no structural differences from the standard attack version of the aircraft

A-10 Warthog (79-0097) of the 354th Tactical Fighter Wing, Myrtle Beach,
North Carolina, during a visit to Andrews AFB, Maryland in about May 1983
(*Robert F Dorr*)

AK tailcode denotes the 18th Tactical Fighter Squadron, part of the 343rd Tactical Fighter Wing at Eielson AFB, Alaska. This A-10 Warthog (80-0254) was visiting the annual airshow at Ontario, Canada in October 1989 (*David F Brown*)

Big friend, little friend. The tremendous disparity in aircraft markings is pointed up by this 10 February 1988 comparison of Warthog 77-0271 (seen elsewhere on these pages in its post-Desert Storm incarnation) and the Beech T-34C Turbo-Mentor (bureau no 162269), both of which are parked in the rain at Andrews AFB, Maryland. Ironically, the Navy's orange/white scheme for its primary trainer would not have made the Warthog any more visible than its 'Europe One' green colours in the deserts of the Middle East (*Robert F Dorr*)